North American
XB-70 Valkyrie
A Photo Chronicle

John M. Campbell
with Garry R. Pape

Schiffer Military/Aviation History
Atglen, PA

ACKNOWLEDGEMENTS

We would like to thank the many individuals who assisted us in the creation of this volume. Without their assistance and encouragement this would not have been possible: Dr. James Crowder, Tinker AFB history office; Dr. Donald Klinko, USAF history office at Ogden, Utah; Steve Pace, fellow aviation historian and writer; Mark Bacon, USAF (Ret.); Wayne Waldron, Wayne Watts, Steve Link, Gary James, John Heyer, Jack Moses, Jon Maguire, Col. Kenneth Wilkerson of the Tinker Flight Test Branch, Mark Copeland, Frederick Johnsen of the history office at Edwards AFB, and the personnel at the Wright-Patterson AFB archives. Thanks also to Dorothy Campbell, Mandy Jo Brown, David Brown, James Brown, Steve Brown, and Barbara Pape.

This book is dedicated to those who designed,
manufactured, and flew the XB-70 Valkyrie.

Book Design by Robert Biondi.

Library of Congress Catalog Number: 95-71865

Printed in China.
ISBN: 0-88740-906-7

We are interested in hearing from authors with book ideas on related topics.

Published by Schiffer Publishing Ltd.
77 Lower Valley Road
Atglen, PA 19310
Please write for a free catalog.
This book may be purchased from the publisher.
Please include $2.95 postage.
Try your bookstore first.

INTRODUCTION: North American Aviation's XB-70 Valkyrie

CONTENTS

World War II was drawing to a conclusion, the Army Air Forces interest in nuclear propulsion systems was starting. From this the highly classified Nuclear Energy for the Propulsion of Aircraft, or more simply, NEPA, program was started. In the early 1950s the Boeing Company was working on Project MX-2145 in conjunction with the Rand Corporation. This project dealt with various types of systems, manned and unmanned, designed to carry "high-yield special weapons." On January 22, 1954, Boeing presented the U.S. Air Force with data indicating the possibility of a bomber aircraft powered by chemically-augmented nuclear powerplants. By March Boeing provided additional data; Lockheed Aircraft Corporation and Convair, under contracts with the Office of Aircraft Nuclear Propulsion, provided similar data at about this same time.

The Air Force was quite interested and General Operational Requirement No. 38 was issued in October 1954 calling for a manned intercontinental bomber to replace the B-52 beginning in 1965. In March 1955 General Operational Requirement Nos. 81 and 82 were issued. Number 81 covered the development of a nuclear-powered strategic weapon system capable of 11,000 nautical mile radius and a mach 2 plus speed capability over 1,000 miles at an altitude of greater than 60,000 feet. Requirement No. 82 was a direct replacement for No. 38. The new requirement stated that the new bomber was to carry a 20,000 lb. payload of nuclear weapons. The execution of these requirements fell upon the USAF's Air Research and Development Command. To implement Requirement No. 82, ARDC issued Study Requirement No. 22 under which the new bomber was known as Weapon System 110A.

The Air Force awarded letter contracts to both Boeing and North American Aviation on November 8, 1955 for the Phase I development of Weapon System 110. After nearly two years of competition, and changes in requirements, North American Aviation, Inc. was announced by the USAF as the selected design

Left: An artist's conception of North American Aviation's XB-70. It was originally speed, altitude, and nuclear propulsion that started the Weapon System 110, later XB-70, design competition.

on December 23, 1957; the actual contract coming on January 2, 1958. Not unlike other weapon systems, the XB-70 suffered from the effects of changes in government priorities, budgets, and new technologies. The drive for chemically-powered aircraft fell out of favor in about 1956 and by 1961 nuclear-powered aircraft projects met the same fate. Manned bombers were thought to be relics of the past as the Intercontinental Ballistic Missile, or ICBM, would be the weapon of the future.

The XB-70A survived the changes of politics and policies, but strictly as an experimental program. The rollout ceremony for the first XB-70A, serial number 62-0001, occurred on May 11, 1964 at Air Force Plant 42 in Palmdale, California. Four and a half months later, on September 21st, North American's Alvin S. White and USAF Col. Joseph F. Cotton took the big white bird into the air for its first flight; a total of one hour and seven minutes were spent airborne on that day. By July 1st, 1965, XB-70A-1 had accumulated 14 flights. On July 17th XB-70A-2, serial number 62-0207, lifted off the ground for the first time; once again Al White and Col. Joe Cotton were at the controls.

Unfortunately, the flight that seems to be remembered by the public was the 46th flight of the second XB-70. On June 8, 1966, it and a number of other aircraft powered by General Electric engines were flying in formation when tragedy struck (this incident is fully covered in Chapter 3). But the design, construction, and performance of these aircraft provided much to the U.S. aircraft industry. In March 1967, the National Aeronautics and Space Administration, or NASA, made its first flight with this aircraft as part of America's supersonic transport (SST) program. Last test flight occurred on December 17, 1968, the 82nd flight for XB-70A-1. The last flight occurred on February 4, 1969 when NASA's Fritz Fulton and USAF's Emil "Ted" Sturmthal flew it to the Air Force Museum at Wright-Patterson AFB near Dayton, Ohio. It remains there today on display.

CHAPTER 1: Development

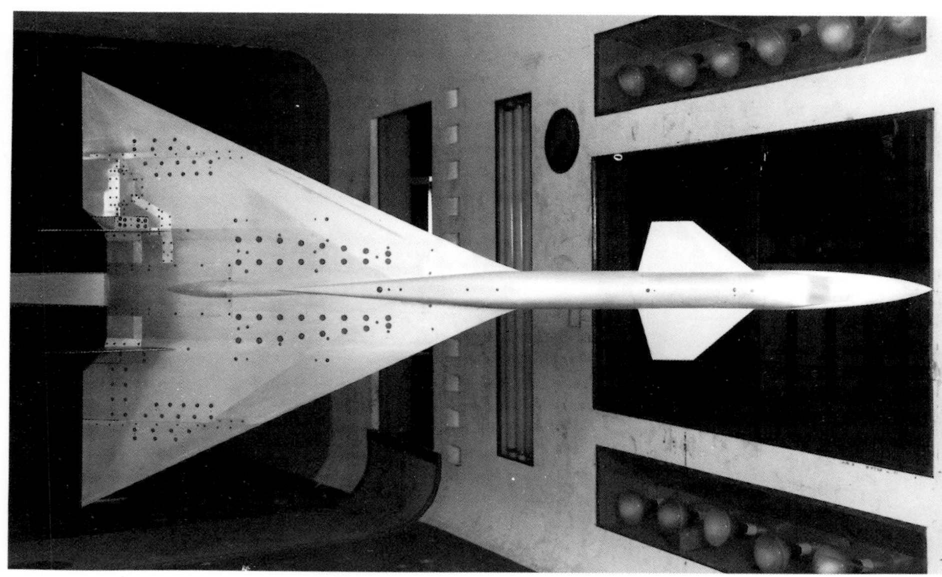

Stainless steel XB-70 wind tunnel model.

Nose section being moved into place to be mated with the rest of the fuselage.

Assembly of the XB-70 took place at North American Aviation-run facilities at Air Force Plant 42 in Palmdale, California. Here the intermediate fuselage is in the process of being moved.

Landing gear retraction/extension tests.

Above and two below: Wing attachment operation.

Above: With wing stubs attached and on its own gear, the gigantic XB-70 moves on its own from one assembly station to another. Below: Tail feathers attached, wings will be next.

Above: Skate trim operation along the wing to wing stub area.

Above right: Moveable wing tip being attached. The workers in this picture help to illustrate the immensity of this huge bomber.

Right: Airframe assembly operations nearing completion.

Installation of the General Electric YJ93-GE-3 engines.

The first XB-70A returning from the paint shop shortly before its official rollout on May 11, 1964. Note that the wing tips are level at zero degrees.

Between roll-out and first flight many ground tests needed to be conducted. Here engine runup tests are being accomplished.

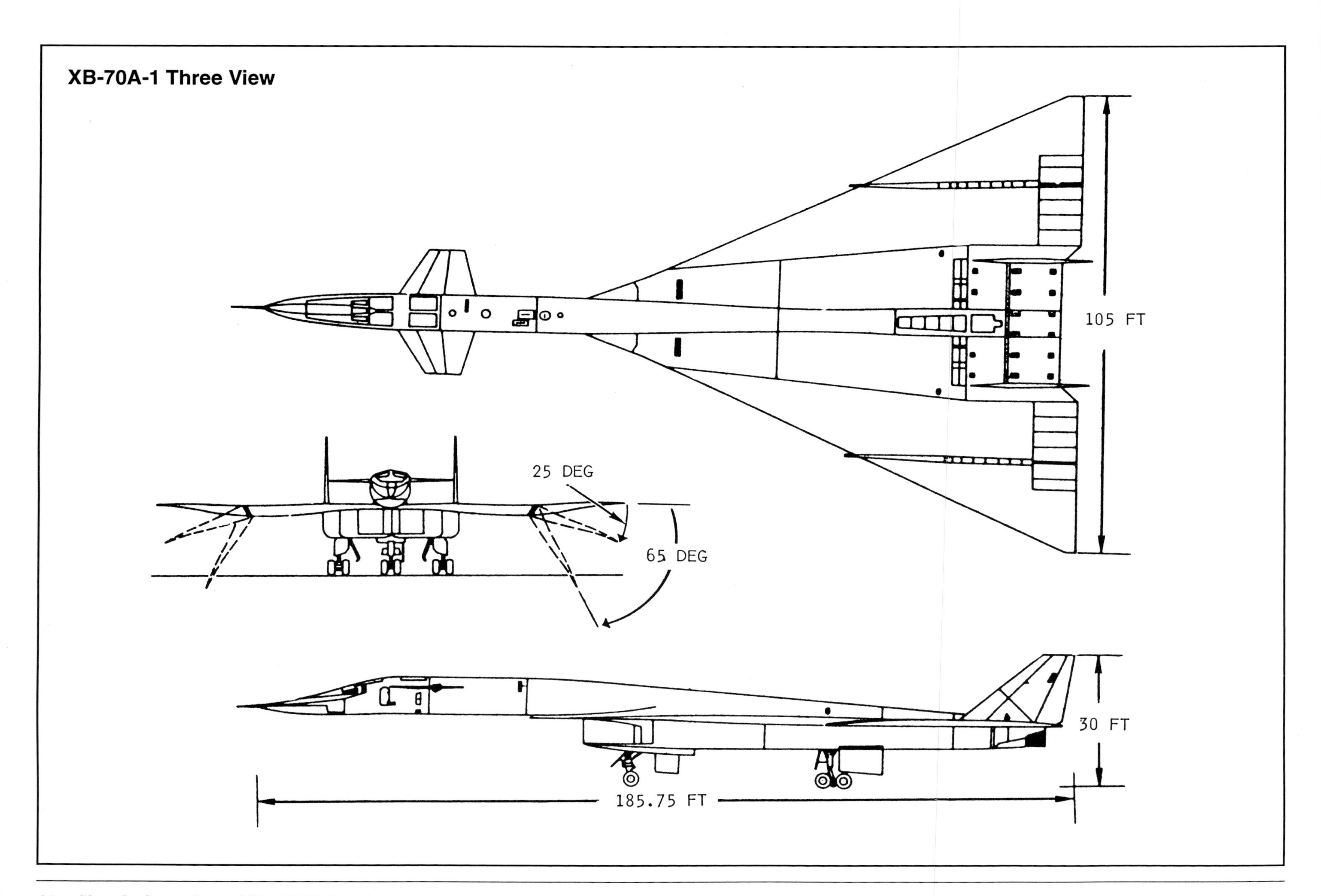
XB-70A-1 Three View
105 FT
25 DEG
65 DEG
30 FT
185.75 FT

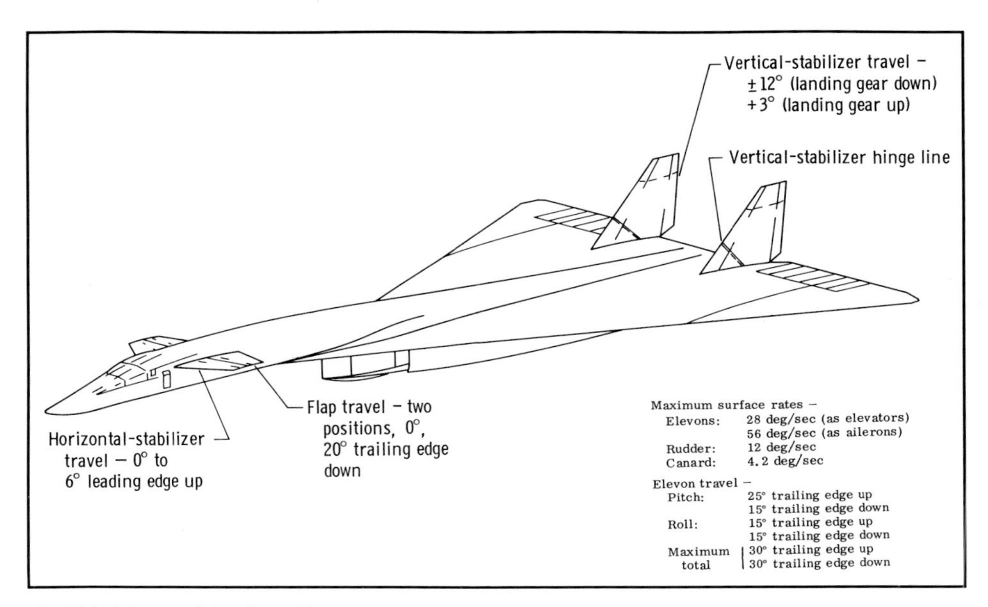

XB-70A-1 Control Surface Movements

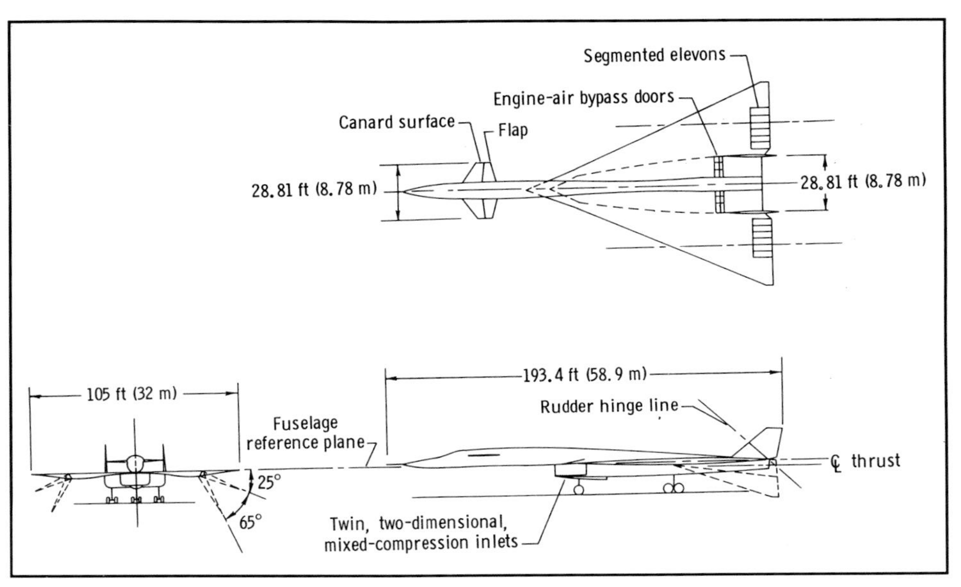

XB-70A-1 Leading Particulars

XB-70A-1 Configuration Data

Wing:
Total Area (Includes 2,482.34 sq. ft. covered by fuselage but not 33.53 sq. ft. of the wing ramp area) .... 6,297.8 sq. ft.
Span .... 105 ft.
Aspect Ratio 1.751
Taper Ratio 0.019
Dihedral Angle .... 0 deg.
Root Chord (At Wing Station 0) .... 117.76 ft.
Tip Chord (At Wing Station 630 in.) .... 2.19 ft.
Mean Aerodynamic Chord .... 942.38 in.
  Wing Station .... 213.85 in.
  Fuselage Station of 25% Wing Mean Aerodynamic Chord .... 1,621.22 in.
Sweepback Angle
  Leading Edge .... 65.57 deg.
  25% Element .... 58.79 deg.
  Trailing Edge .... 0 deg.
Airfoil Section .... 0.30 to 0.70 Hex (Mod)
Thickness (percent chord):
  Wing Station
  Root to 186 in. 2.0
  460 in. to 630 in. 2.5
Folding Wing Tip (per tip)
  Area .... 520.90 sq. ft.
  Span .... 20.78 ft.
  Aspect Ratio 0.829
  Taper Ratio 0.046
  Root Chord (At Wing Station 380.62 in.) .... 47.94 ft.
  Tip Chord (At Wing Station 630 in.) .... 2.19 ft.
  Mean Aerodynamic Chord (At Wing Station 467.37 in.) .... 384.25 in.
  Down Deflection .... 0, 25, and 65 deg.
  (From Inboard Wing)
Elevons (per elevon)
  Total Effective Area Aft of Hinge Line ... 197.7 sq. ft.
  (Includes 3.33 sq. ft. air gap at wing tip fold line)
  Span ....
  Wing Tips Up .... 20.44 ft.
  Wing Tips Down .... 13.98 ft.
  Chord .... 116 in.
  Sweepback of Hinge Line .... 0 deg.
Canard
  Area (Including 150.31 sq. ft. Covered by Fuselage) .... 415.59 sq. ft.
  Span .... 28.81 ft.
  Aspect Ratio 1.997
  Taper Ratio 0.388
  Dihedral Angle .... 0 deg.
  Root Chord (At Canard Station 0) .... 20.79 ft.
  Tip Chord (At Canard Station 172.86 in.) .... 8.06 ft.
  Mean Aerodynamic Chord .... 184.3 in.
  Canard Station .... 73.71 in.
  Fuselage Station of 25% Chord .... 553.73 in.
  Sweepback Angle
  Leading Edge .... 31.70 deg.
  25% Element .... 21.64 deg.
  Trailing Edge .... -14.91 deg.
  Airfoil Section .... 0.34 to 0.66 Hex (Mod)
  Thickness Chord Ratio
  Root 2.5
  Tip 2.52
  Ratio of Canard Area to Wing Area 0.066
Canard Flap (per side)
  Area (Aft of Hinge Line) .... 54.69 sq. ft.
  Inboard Cord (At Canard Station 47.93 in.) .. 7.16 ft.
  Outboard Chord (At Canard Station 172.86 in.) .... 3.34 ft.
  Ratio of Flap Area to Canard Semiarea 0.263
Vertical Tail (per vertical)
  Area (Includes 8.96 sq. ft. Blanketed Area) .... 233.96 sq. ft.
  Span .... 15 ft.
  Aspect Ratio 1
  Root Chord (At Vertical Tail Station 0) .... 23.08 ft.
  Tip Chord (At Vertical Tail Station 180 in.) .... 6.92 ft.
  Taper Ratio 0.30
  Mean Aerodynamic Chord .... 197.40 in.
  Vertical Tail Station .... 73.85 in.
  Fuselage Station of 25% Chord .... 2,188.50 in.
  Sweepback Angle
  Leading Edge .... 51.77 deg.
  25% Element .... 45 deg.
  Trailing Edge .... 10.89 deg.
  Airfoil Section .... 0.30 to 0.70 Hex (Mod)
  Thickness Chord Ratio
  Root 3.75
  Tip 2.50
  Cant Angle .... 0 deg.
  Ratio of Vertical Tail to Wing Area 0.037
Rudder
  Area (Includes 8.66 sq. ft. Blanketed Area) .... 191.11 sq. ft.
  Span .... 15.00 ft.
  Root Chord (At Vertical Tail Station 0) .... 9.16 ft.
  Tip Chord (At Vertical Tail Station 180 in.) .... 6.92 ft.
  Sweepback of Hinge Line .... -45.0 deg.
  Ratio of Rudder Area to Vertical Tail Area 0.82
Fuselage (Including Canopy)
  Length .... 185.75 ft.
  Maximum Depth (At Fuselage Station 878 in.) .... 106.92 in.
  Maximum Breadth (At Fuselage Station 855 in.) .... 100 in.
  Side Area .... 939.72 sq. ft.
  Planform Area .... 1,184.78 sq. ft.

CHAPTER 2: Test Flight

Ground crewmen, busy about their duties, had their hands full with the XB-70. Apex of the XB-70A delta wing, which swept more than the main wing, provided the most efficient recovery of the air into the inlet. The rectangular inlet ducts were assisted by the fuselage boundary layer on the underside of the wing apex.

The XB-70 had a windshield that moved along a variable nose ramp. Like SST aircraft, the forward cockpit could be lowered for subsonic speeds to increase visibility. The forward canard slabs provided trim control and reduced drag during all aspects of flight.

The XB-70, serial number 62-0207, awaiting another flight is seen hooked up to the auxiliary power units and other equipment. With wing tips at zero degrees and ready to soar, the picture is quite serene in this view. The number one ship is seen in the distance.

The XB-70A-1 just seconds into her maiden flight after rotating at 4,853 feet. This flight on September 21, 1964 only lasted 107 minutes. Piloted by Al White and Joe Cotton from Palmdale, California, to nearby Edwards AFB, both rated the XB-70 as flying quite well.

With emergency vehicles in tow, the XB-70A-1 taxis out for yet another of the many tests prescribed for the XB-70 project.

Above: The XB-70, in level flight with wing tips set at about 16 degrees, flies over California's Mojave Desert. To observe all aspects of the flight test, a chase plane flies along side.

Opposite: This grouping of nine pictures shows frontal and side views as seen by the flight test photographer. In the upper and lower series of three photos, the Convair B-58 Hustler is seen in the role of chase plane.

This photo of the XB-70 and accompanying B-58 reveals a nice view from above. The downward slant of the wing tips is quite visible from this viewpoint.

This sequence of six frames reveals a graceful climb into the warm California skies. The three lower images show how lean and sleek the XB-70 appears in takeoff posture.

Like a hooded cobra rising to strike, the XB-70 settles in for another graceful landing. The massive area of the wing cannot displace the air, and moisture in the air causes the smoke from the wheels to cascade over the aft control surfaces.

In this sequence of six images, we see the different positions of the adjustable wing tips.

This series of four photos depicts different views while taxiing in after another test hop. The lower left shows the parachute drogue fully deployed to assist in a quicker reduction in ground speed.

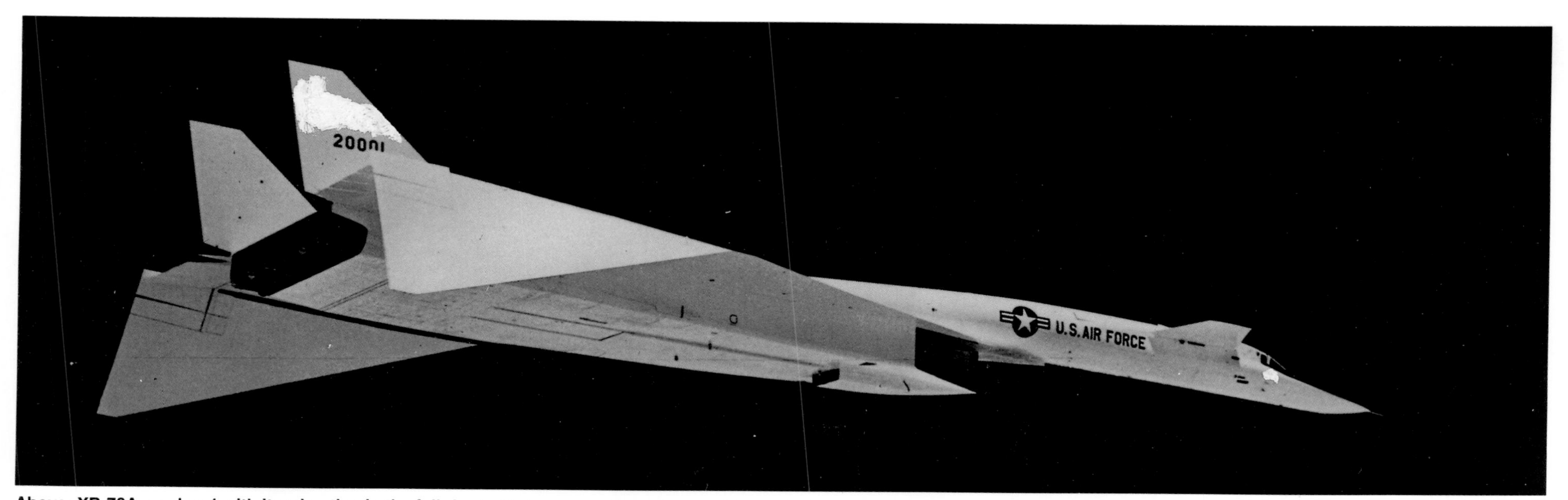

Above: XB-70A number 1 with its wing tips in the full-down 65 degree configuration. In this particular mode of flight, wing tips offered high-speed stability, both lateral and longitudinal, and a five percent increase in lift. Below: XB-70A number 1 on its seventh flight on March 4, 1965. Wing tip settings are seen here at 25 percent.

Northrop's T-38 Talon is seen here slightly above the XB-70 in the role of chase plane. The XB-70 shows an impressive clear view of its underbelly and clean lines.

The final landing stages of the XB-70 required the use of three full-size parachutes.

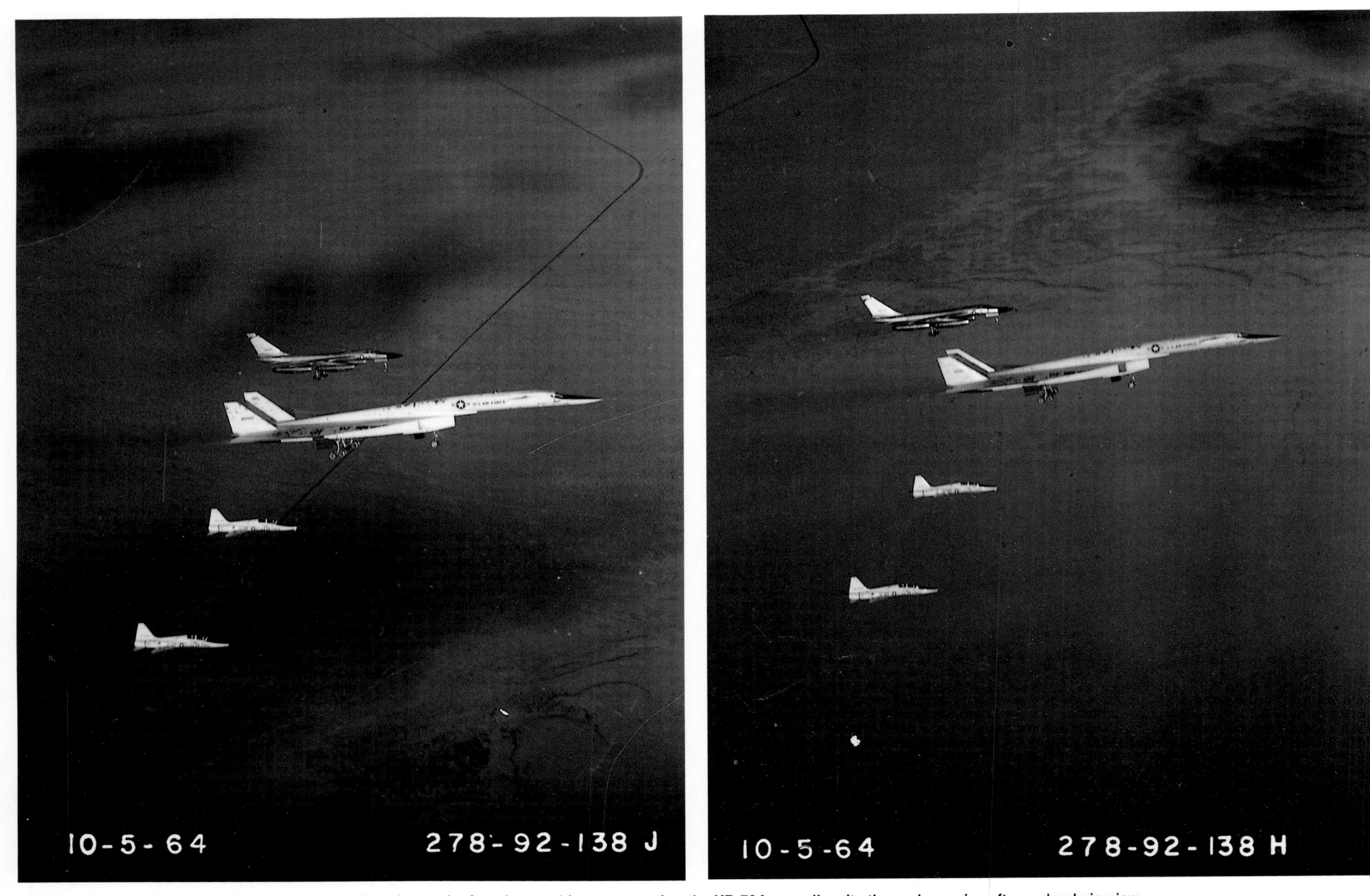

These two views were special for the photographer, due to the fact that on this rare occasion the XB-70A as well as its three chase aircraft are clearly in view.

Jacks in place and secured, the number two aircraft prepares to undergo landing gear functional tests as well as weight and balance requirements.

The XB-70 2,000 MPH research aircraft and the X-15A-2 hypersonic research vehicle built by North American Aviation, Inc. Los Angeles Division's facility in Palmdale, California.

This view of the XB-70 as it sits outside the hangar shows a good aspect of the markings and details.

During a ground engine test, the raw power of the XB-70 is clearly seen here with all six engines burning brightly.

At 10:57 a.m. on February 4, 1969, the sole remaining XB-70 rotated and climbed away from what had been its home for over five years, Edwards Air Force Base, California. Her destination — Wright-Patterson AFB, Ohio, and home of the Air Force Museum.

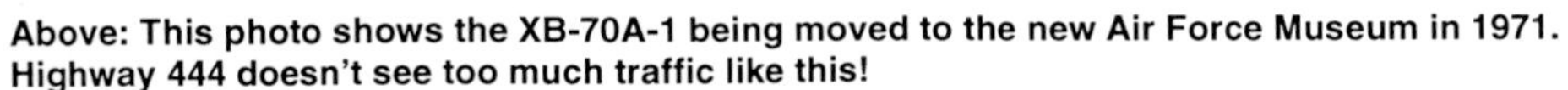

Above: This photo shows the XB-70A-1 being moved to the new Air Force Museum in 1971. Highway 444 doesn't see too much traffic like this!

Above right: The XB-70A-1 after arriving at Wright-Patterson AFB. She sits berthed in front of a German Junkers Ju-88 of World War II, a very interesting contrast since they were manufactured barely 30 years apart.

Right: Left to Right: Guy Strow, Col. Fritzhugh L. Fulton, Col. Roscoe Turner, Alvin S. White, Walt Spivak, Van H. Shepard, and Col. Joseph F. Cotton. Fulton flew the XB-70 31 times as a pilot and 32 times as copilot for the USAF and NASA; White flew it 49 times as pilot and 18 as copilot for NAA; Shepard flew it for NAA 23 times each as pilot and copilot; and Cotton was credited with 19 flights as pilot and 43 as copilot for the USAF.

Above left: North American Aviation president Warren Swanson presents the Mach 3 award to Van Shepard. Al White and Shepard made the highest flight in the XB-70 on March 19, 1966 when they reached 74,000 feet.

Above: North American's Al White, on the left, and Col. Joe Cotton in the first XB-70 on September 21, 1964 and the fastest flight, Mach 3.08, on April 12, 1966. As the news media presses in on them, General Manson, in the center, congratulates the two test pilots.

Left: Col. Joe Cotton, on the left, and NASA's Donald L. Mallick. Col. Cotton flew on the first flight of both XB-70s. Don Mallick flew the XB-70 four times as its pilot and five times in the copilot's seat.

CHAPTER 3: XB-70 Accident

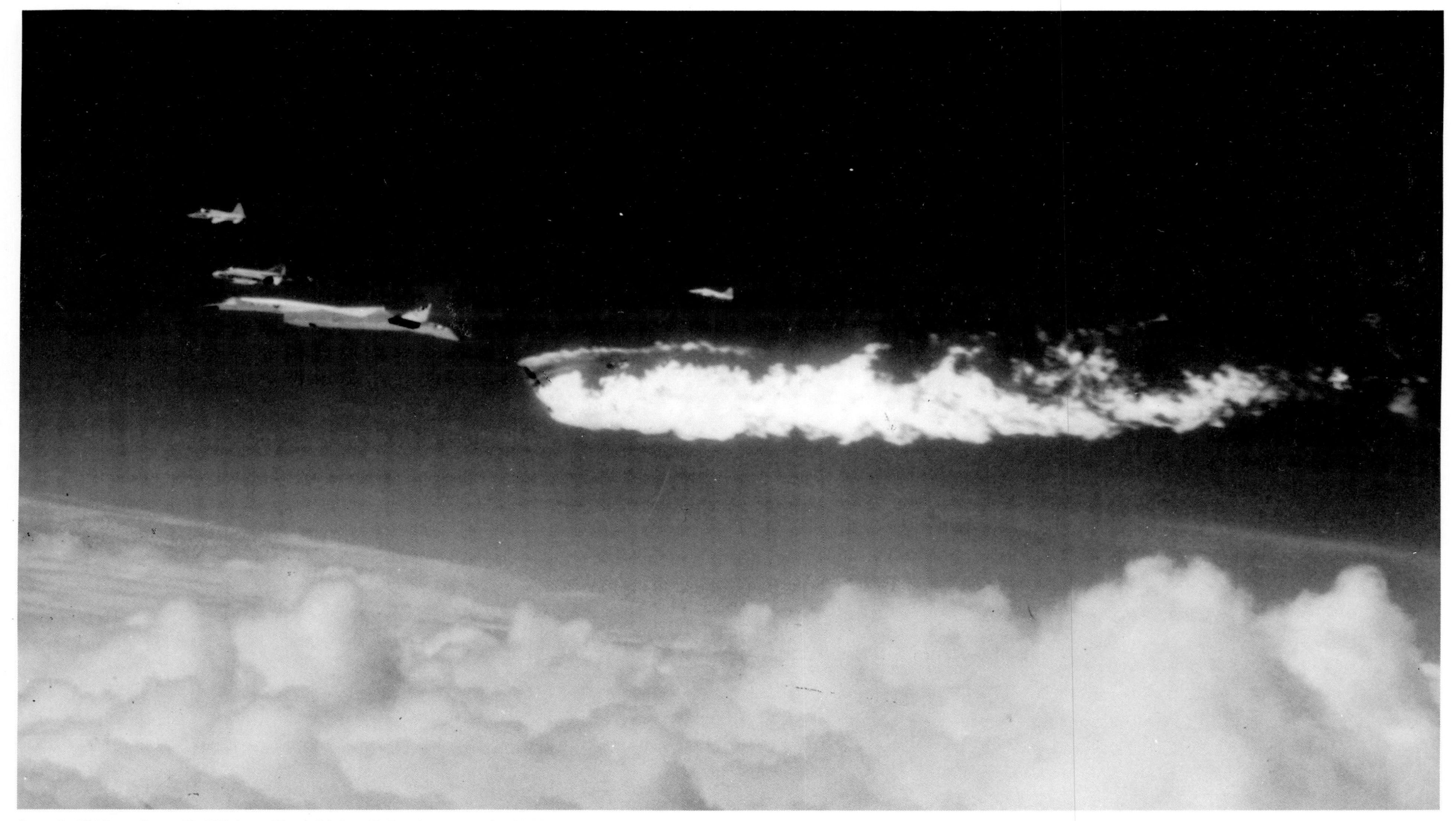

June 8, 1966 at about 25,000 feet. The initial collision between the F-104 and the XB-70A-2 was most likely caused by unperceived closure between the F-104's empennage and the XB-70's wing tip, which was out of view of NASA pilot Joe Walker who was flying the fighter. Both aircraft were functioning normally up to the point of impact. Walker was fatally injured when his F-104 hit the tailplane of the XB-70.

The XB-70 is seen here in its final roll, going into an irreversible flat spin which ultimately ended in its destruction.

Without sufficient vertical tail area left to hold the XB-70 straight and level, it rolled over on its back and with nose down went into a violent yaw. It was at this moment that pilot Al White of North American and copilot Maj. Carl Cross (USAF) realized that their craft was doomed.

Seen from high above the desert floor, the luminous plume of smoke rises as a quiet reminder of the high risks involved in the flight test arena (and of photo opportunities).

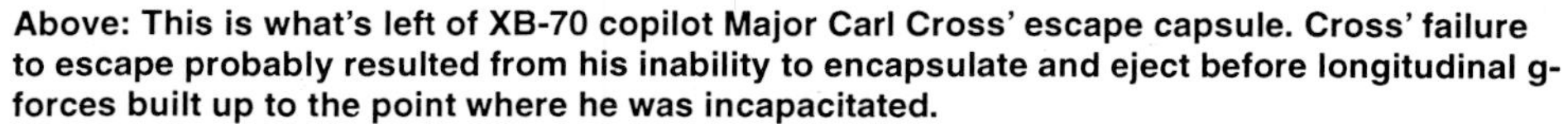

Above: This is what's left of XB-70 copilot Major Carl Cross' escape capsule. Cross' failure to escape probably resulted from his inability to encapsulate and eject before longitudinal g-forces built up to the point where he was incapacitated.

Above right: The XB-70 entered into a flat spin, hit the desert floor in a relatively flat attitude, and burned.

Right: The mid-air collision between the XB-70 and the F-104 resulted in the fatal injuries of Walker and Cross. In this view is seen the aft section of the F-104 chase plane.

Above: Al White received minor injuries when his right arm became jammed between the ejection hand grip and the edge of the upper escape capsule door. He also received additional minor injuries when the capsule hit the ground with the impact attenuator bladder uninflated.

Right: Rear view of Al White's capsule as it rests on the desert floor. Both escape capsules would have functioned normally if they had been operated as intended.

CHAPTER 4: XB-70 In Detail

This is an excellent view of the emergency egress totally exposed. The actuator handles and safety harness are still the same generic type as used in the military's more conventional aircraft. North American Aviation's test pilot Al White poses here with the module.

In this view of the capsule, we see the module in its half closed position. This reveals the protection offered the flight crewman.

This image clearly defines the lines and apparatus of the fully closed capsule as seen here being installed into the pilot's station in the XB-70.

Right: The XB-70 escape pod is seen here during its test firing which was ground launched from a position of no forward motion.

Pilots Al White, on the left, and Joe Cotton, on the right, in the cockpit of the XB-70. The cockpit area from the rear appears to be quite clean and free of the massive array of instrumentation that one would expect to see in an aircraft of this size. The main instruments are, however, hidden by the quite sizeable escape pods.

This view shows a close-up of the throttle quadrants, TACAN, radio, hydraulic readouts, nose steering, yaw trim, and other assorted controls all assembled within the centerline of the cockpit.

The cockpit section of XB-70A-1 as seen here with a straight on view shows the throttle quadrant centered between the pilots' station and the control yokes, as well as the individual pressure, temperature, and RPM gauges.

This view of the copilot's station shows the control wheel and fuel tank sequence positioned on the left of the control wheel. Fuel gauges are also visible here.

This aft end view of the XB-70 shows the six YJ93-3 engines. For Mach 3 speeds, diverging control flaps would be fully open as well as for takeoff. The flaps are shown here in near closed position.

The XB-70 undergoing drop check of the landing gear system. The gear retracts backward, the carriage jackknifes and the entire unit twists prior to closing. The brake discs are mounted between the main gear wheels.

This view of the main gear shows the gear as it was retracted backwards just prior to stowing in the wheel well.

Above: The XB-70 seen here just behind one of its six mighty General Electric YJ93 turbojet engines. On May 19, 1966 the XB-70 sustained mach 3 flight for 32 minutes. Maximum speed in level flight at sea level was 685 mph. Normal range was 2,650 miles. Below: The XB-70, seen here in the hangar at North American undergoing static structural load testing.

Above: During the late 1960s and 1970s most of the major aircraft manufacturers were contemplating supersonic transport (SST) designs. These aircraft were to be used as passenger aircraft and speed up air transportation. The XB-70A seen here is fitted with window templates near mid fuselage and was part of North American's SST program. Below: Here weight and balance tests are being conducted on the XB-70A in a hangar at Edwards AFB.